# The Casio SL-450
## A TOOL FOR TEACHING M

# The Casio SL-450

## A TOOL FOR TEACHING MATHEMATICS

### DAVID J. GLATZER & JOYCE GLATZER

CONSULTING EDITOR: MAX SOBEL

**CASIO, INC.** • DOVER, NEW JERSEY

© 1993 by CASIO, Inc.
570 Mt. Pleasant Avenue
Dover, New Jersey 07801

The contents of this book can be used by the classroom teacher to make reproductions for student use. All rights reserved. No part of this book may be reproduced or utilized in any form or by any means, electronic or mechanical, including phtocopying, recording or by any information storage or retrieval system without permission in writing from CASIO, Inc.

Printed in the United States of America

ISBN 1-878532-05-7

# Contents

| Activity | Grade | Topic | Page No. |
|---|---|---|---|
| **Foreword** | | | vii |
| **Key Operation Summary** | | | viii |
| **Quick Start Reference** | | | ix |
| **Forward Count** | K-1 | *counting on* | 1-2 |
| **Again and Again** | 1-3 | *skip counting* | 3-4 |
| **In Reverse** | 1-3 | *counting back* | 5-6 |
| **Knowing Your Place** | 2-3 | *place value* | 7-8 |
| **Read My Mind** | K-3 | *basic facts* | 9-10 |
| **Squeeze Play** | 2-3 | *estimating sums, differences* | 11-12 |
| **Pitch It In** | 1-3 | *problem solving* | 13-14 |
| **Making Cents** | 1-3 | *problem solving* | 15-16 |
| **Target Practice** | 4-5 | *estimating products* | 17-18 |
| **Leftover Crunchies** | 4-5 | *remainders in division* | 19-20 |
| **Point The Way!** | 4-5 | *decimal sums, differences* | 21-22 |
| **What's The Point?** | 5-6 | *decimal products* | 23-24 |
| **Don't Touch That Key!** | 4-6 | *problem solving* | 25-26 |
| **Can You Top This?** | 4-5 | *problem solving* | 27-28 |
| **The Race Is On** | 4-6 | *mental math* | 29-30 |
| **First Class Mail** | 4-6 | *problem solving* | 31-32 |
| **Mean Machine** | 4-6 | *finding averages* | 33-34 |
| **Name That Part** | 4-6 | *fractions* | 35-36 |
| **Expand Your Memory** | 3-6 | *problem solving* | 37-38 |
| **Key Purchases** | 4-6 | *problem solving* | 39-40 |
| **Putting Things in Order** | 5-6 | *operational order* | 41-42 |
| **Power Up** | 6 | *exponents* | 43-44 |
| **Exit Sign** | 5-6 | *integers* | 45-46 |
| **Follow The Leader** | 1-3 | *patterns* | 47-48 |
| **Next In Line** | 4-6 | *patterns* | 49-50 |
| **Short Cut** | 5-6 | *patterns* | 51-52 |
| **Family Affair** | 5-6 | *percent of numbers* | 53-54 |
| **Making the Grade** | 5-6 | *ratios to percent* | 55-56 |
| **The Price is Right** | 6 | *add-ons & discounts* | 57-58 |
| **Square Deal** | 6 | *area of square* | 59-60 |

# Foreword

*The Curriculum and Evaluation Standards for School Mathematics,* published by the National Council of Teachers of Mathematics states: "Calculators must be accepted at all levels as a valuable tool for learning mathematics. Calculators enable children to explore number ideas and patterns, to have valuable concept-development experiences, to focus on problem solving processes, and to investigate realistic applications. The thoughtful use of calculators can ensure the quality of the curriculum as well as the quality of children's learning."

Consistent with that position, this book offers to classroom teachers, grades K-6, activities that allow the student to enhance their study of mathematics by using the calculator to explore, develop, practice, and extend mathematical concepts.

This book is more than just a "how to" use the calculator book. It contains teaching lessons that allow the student to gain competence with the SL-450, while concurrently learning mathematics.

The activities are identified by topic and delineate grade level use. Among the topics included are: numeration, place value, estimation, whole number operations, decimals, integers, exponents, percent, operational order, geometry & measurement, fractions, and patterns. It should be noted that while grade levels are given, each activity has the potential to be adapted for use with other grades. The teacher notes contain many suggested questions that support the activity and enable students to make connections between the activity and its related mathematical concept. Extensions, which often include non-calculator activities, are suggested where appropriate to enrich and extend the student's learning. Each activity includes a problem titled "Thinking Cap" which requires higher-level thinking skills to complete. The "Thinking Cap" is related to the type of activity presented and demands different degrees of calculator involvement.

In keeping with the Standards, problem solving, reasoning, connecting, and communicating with and about mathematical concepts is integrated into each activity. As a result, students should find that effective use of the calculator requires them to know more mathematics, not less.

The use of the calculator opens the door to exploration and discovery in mathematics for all students, especially those for whom computational deficiencies have limited their experiences. Making the student an informed consumer of the calculator, requiring them to know when it is appropriate to use a calculator and when it is not, is the responsibility of all educators.

# Key Operation Summary

**[AC]** Turns the calculator on. Clears memory and display.

**[C]** Clears the last entry in a pending operation.
22 [+] 63 [C] 65 [=] *displays* 87

**[.]** Enters a decimal point in a number.
[.] 546 *displays* 0.546

**[=]** Completes a calculation.
22 [+] 63 [=] *displays* 85

Functions as a constant operation.
2 [+] [+] [=] [=] [=] [=] continually adds 2 to number in display
2 [−] [−] [=] [=] [=] [=] continually subtracts 2 from number in display
2 [x] [x] [=] [=] [=] [=] continually multiplies number in display by 2
2 [÷] [÷] [=] [=] [=] [=] continually divides number in display by 2.

**[+]** Adds whole numbers, decimals, integers.
23 [+] 65 [=] *displays* 85

**[−]** Subtracts whole numbers, decimals, integers.
65 [−] 23 [=] *displays* 42

**[x]** Multiplies whole numbers, decimals, integers.
23 [x] 65 [=] *displays* 1495

**[÷]** Divides whole numbers, decimals, integers.
144 [÷] 6 [=] *displays* 24

**[%]** Calculates percent.
25% of 48: 48 [x] 25 [%] *displays* 12
$48 plus 6% tax: 48 [x] 6 [%] [+] *displays* 50.88
$48 less 25% discount: 48 [x] 25 [%] [−] *displays* 36
What % of 25 is 4?: 4 [÷] 25 [%] *displays* 16

**[√]** Calculates square root of a number.
25 [√] *displays* 5

**[+/−]** Changes the sign of a number.
5 [+/−] *displays* -5
5 [+/−] [+/−] *displays* 5

**[M+]** Adds the displayed value to the value in memory.

**[M−]** Subtracts the displayed value from the value in memory.

**[MR]** Displays contents of the memory.
23 [x] 45 [=] *displays* 1035
[M+] [C] [MR] *displays* 1035
[C] 65 [M+] [MR] *displays* 1100
65 [M−] [MR] *displays* 0

The Casio SL-450: A Tool for Teaching Mathematics

# Quick Start Reference

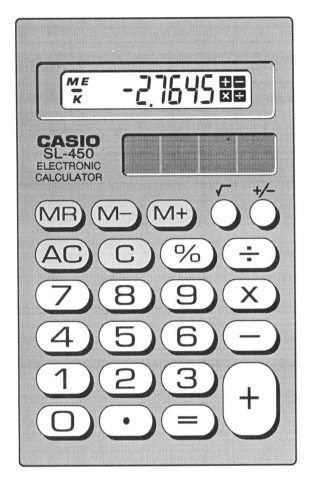

|  |  |
|---|---|
| AC | used to turn calculator on (clears memory and display) |
| 0 - 9 | used to enter digits 0-9 |
| . | decimal point |
| C | clears the last entry |
| + | adds next number to number in display |
| − | subtracts next number from number in display |
| x | multiplies the number by the number in the display |
| ÷ | divides the number in display by next number entered |
| = | completes operations and displays results |
| +/− | changes the sign of the number in display |
| % | used to calculate percent |
| √ | calculates square root of number in display |
| M+ | adds number in display to the memory |
| M− | subtracts number in display from memory |
| MR | displays contents of memory |

## Special Functions available: CONSTANTS

*constant* [+] [+] *number* [=] ...  automatic constant for [+]. Continually adds constant to the number in the display.

*number* [−] [−] [=] ...  automatic constant for [−]. Continually subtracts number from number in the display.

*number* [x] [x] [=] ...  automatic constant for [x]. Continually multiplies the number times number in the display.

*number* [÷] [÷] [=] ...  automatic constant for [÷]. Continually divides number in the display by the number.

**The Casio SL-450: A Tool for Teaching Mathematics**

# Forward Count

Name_____

## Problem 1:

Enter [AC] [1] [+] [+]   Display: K  1. ⊞

[=]

[=]

[=]

[=]

[=]

What is the calculator doing?

## Problem 2:

Enter [AC] [1] [+] [+] [7] [=]   Display: 8.

[=]

[=]

[=]

What is the calculator doing?_____

Where did it start?_____

## Thinking Cap

What number would you enter for ? to get a display of 25?  Complete the displays.

Enter [AC] [1] [+] [+] [?]   Display: 25.

[=]

[=]

[=]

© 1993 CASIO, INC.                    **The Casio SL-450: A Tool for Teaching Mathematics**    **1**

# TEACHER NOTES: *Forward Count*

**Objective:** To count on by one.

**Grade Level: K-1**

**Topic:** *Numeration*

## Using the Activity:

This activity can be completed with students using calculators or can be done with the teacher using the overhead version of the SL-450 (model OH-450) and students responding to the display. Entering **number** [+] [+] activates the automatic constant for addition on the SL-450. This results in the indicated number being added again and again to the display as equal is pressed. Students should count orally as the equal key is pressed. Students could be asked to predict the next number before the equal key is pressed to check for understanding of number sequence.

What is the calculator doing? *counting on from 1*

## In problem 1, ask students:

How many times do you have to press the equal key to display 18? answers will vary *(17)* Why? *(18 is 17 more than 1)*

Take their response and press [=] that number of times to see what will happen. Were they correct? If not, have them modify their guess.

Reinforce by asking:

How many times do you have to press the equal key to display 39? *(38)*

**For problem 2**, the calculator is counting on by 1's from 7.

**Ask:** How many times did you press [=] to display 11? *(4)* Why? *(11 is 4 more than 7)*

How many times will you have to press the equal key to display 20? *(13)*

Children should have access to a hundred chart to facilitate answering the questions. This will allow students to recognize relationships between numbers.

## Thinking Cap

This section provides a challenge and extension for the learner. Children suggest numbers for the question mark before completing the sequence using a calculator.

## Extension:

Enter [1] [+] [+]

Close your eyes. Press the equal key over and over until you think the calculator displays 20. Stop. Open your eyes. Were you right? How did you know when to stop? How much time will it take to reach 100? 1000?

# Again and Again!    Name_____

Enter the given key sequence into your calculator. Color over each number displayed in charts below as [=] is pressed again and again.

## Problem 1:

Press [AC]  Enter 2 [+] [+] 0 [=] [=] [=] [=] . . .

|  |  |  |  |  |  |  |  |  |  |
|---|---|---|---|---|---|---|---|---|---|
| 1 | 2 | 3 | 4 | 5 | 6 | 7 | 8 | 9 | 10 |
| 11 | 12 | 13 | 14 | 15 | 16 | 17 | 18 | 19 | 20 |
| 21 | 22 | 23 | 24 | 25 | 26 | 27 | 28 | 29 | 30 |
| 31 | 32 | 33 | 34 | 35 | 36 | 37 | 38 | 39 | 40 |
| 41 | 42 | 43 | 44 | 45 | 46 | 47 | 48 | 49 | 50 |

## Problem 2:

Clear the calculator. Press [AC].

Enter 3 [+] [+] 0 [=] [=] [=] [=] . . .

|  |  |  |  |  |  |  |  |  |  |
|---|---|---|---|---|---|---|---|---|---|
| 1 | 2 | 3 | 4 | 5 | 6 | 7 | 8 | 9 | 10 |
| 11 | 12 | 13 | 14 | 15 | 16 | 17 | 18 | 19 | 20 |
| 21 | 22 | 23 | 24 | 25 | 26 | 27 | 28 | 29 | 30 |
| 31 | 32 | 33 | 34 | 35 | 36 | 37 | 38 | 39 | 40 |
| 41 | 42 | 43 | 44 | 45 | 46 | 47 | 48 | 49 | 50 |

Which numbers are colored on both charts? _____

© 1993 CASIO, INC.                    The Casio SL-450: A Tool for Teaching Mathematics

# TEACHER NOTES: *Again and Again!*

**Objective:** To skip count by a given number.

**Grade Level:** 1-3

**Topic:** *Numeration/Operation Readiness (x)*

## Using the Activity:

This activity uses the automatic constant for multiplication to develop the concept of multiplication. Students should work in pairs, one student enters the data into the calculator and calls out the display, while the other child colors the numbers displayed in the chart. Roles are reversed for the second part of the sheet.

**For problem 1,** students must key in 2 [+] [+] 0 to activate the automatic constant and start counting at 0. This will result in a direct correlation between the number of times the equal key is pressed and the resulting multiplication relationship.(Note: 2 ++=== gives 8 not 6)

2 [+] [+] 0 [=] [=] [=] gives 6, equating to

2 + 2 + 2   (3 additions of 2) which also gives 6, equating to

3 x 2  (3 times 2) which is 6

## Ask students:

1. *What is the calculator doing?*  skip counting by 2
2. *If you press [=] seven times what number is displayed?*  14
3. *What number do you think will be displayed next?*  16 is displayed after 14
4. *How many times do you think you have to press [=] to display 24?*  answers will vary  (12 )
   Try it!  Were you right?
5. *If the chart were extended, do you think 65 would be displayed?  Why or why not?*  No.  Only even numbers are displayed when counting by 2 beginning at 0.

Repeat the questions for **problem 2**.

Children should see that 6, 12, 18, 24, 30, 36, 42, & 48 are displayed on both charts.

Continue the activity each time beginning with a different number, such as 4 [+] [+] [=] or 5 [+] [+] [=].  Make comparisons after each exploration.

## Thinking Cap

This section has students use estimation skills as well as patterns to predict that [=] will have to be pressed 50 times to display 100. Encourage students to use the guess and check problem-solving strategy to find the answer.

## Extension

Use manipulatives and a number line to show skip counting concretely.

# In Reverse

Name_____

Enter the following into your calculator. Record each number displayed.

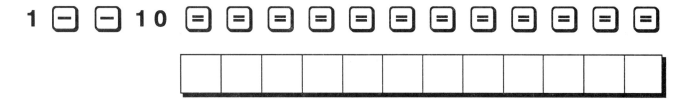

|   |   |   |   |   |   |   |   |   |   |   |   |   |
|---|---|---|---|---|---|---|---|---|---|---|---|---|

What is the calculator doing?_____

**Clear the calculator. Enter.**

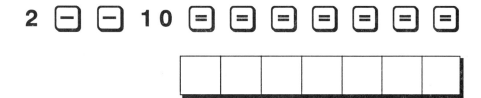

|   |   |   |   |   |   |   |
|---|---|---|---|---|---|---|

What is the calculator doing?_____

How many times did you press ⊟ to get 0?_____

**Clear the calculator. Enter.**

3 ⊟ ⊟ 10 ⊟ ⊟ ⊟ ⊟ ⊟ ⊟

|   |   |   |   |   |   |
|---|---|---|---|---|---|

Did the calculator display 0?_____Explain_____

_____

## Thinking Cap

There are several numbers which can be subtracted over and over again from 20 until 0 is displayed on the calculator. Find all the numbers that work.

© 1993 CASIO, INC.

# TEACHER NOTES: *In Reverse*

**Objective:** To count back by a given number.

**Grade Level:** 1-3

**Topic:** *Numeration/Operation Readiness (Division)*

## Using the Activity:

This activity uses the automatic constant for subtraction to develop the concept of division. Have students enter the given key sequence into the calculator. As the numbers are displayed on the calculator, students should record the numbers in the boxes provided. The students should see that in the first section the calculator is counting back by 1's, while in the second section, it is counting back by 2's.

Ask students:

When the calculator was counting back by 1's from 10, how many times did you press equal to display 7? *(3)* to display 0? *(10)*

What did the calculator display after 0? *(–1)*

When the calculator was counting back by 2's from 10, how many times did you press equal to display 6? *(2)* 0? *(5)*

If you keyed in 5 [–] [–] 10, how many times would you have to press equal to get 0? *(2)*

How come when 3 is repeatedly subtracted from 10, 0 is not displayed? *(10 cannot be divided evenly into groups of 3)*

How many groups of 3 can be subtracted from 10 ? *(3)*

How many items are left ungrouped? *(1)*

## Thinking Cap

Students may use the guess and check problem-solving strategy to discover the numbers that can be repeatedly subtracted from 20 to get a display of 0. These numbers *(20, 10, 5, 4, 2, 1)* represent the whole number divisors of 20 and can be tied to the development of basic division facts.

## Extension

Working in groups of 3, one student uses a calculator, one records the results on paper and one child uses interlocking cubes. The child starts by building a train of 10 interlocking cubes. As equal is pressed, the groups of cubes equal in length to the number being repeatedly subtracted are broken off. Students should see that the length of the group is the number being subtracted, and the number of groups broken off is equal to the number of times the equal key is pressed.

Ex: [2][–][–][1][0][=] [ ][ ][ ][ ][ ][ ][▓]   [2][–][–][1][0][=][=] [ ][ ][ ][ ][ ][ ][▓]

**6** The Casio SL-450: A Tool for Teaching Mathematics

# Knowing Your Place

Name_____

**Enter:** three hundred sixty-nine

**Display:**

In one operation, remove the "6" from the number without changing any of the other digits.

**Display:**

What did you enter?_____ Why?_____

In one operation, change the "3" to a "5" without changing any of the other digits.

**Display:**

What did you enter?_____ Why?_____

| ENTER | CHANGE TO | WHAT DID YOU DO? |
|---|---|---|
| 1. twenty-seven | 7 | |
| 2. sixty-three | 60 | |
| 3. three hundred eighty-five | 85 | |
| 4. four hundred sixty-four | 404 | |
| 5. nine hundred seventy-nine | 970 | |
| 6. three hundred eighteen | 358 | |
| 7. six hundred sixty-four | 964 | |
| 8. one hundred five | 125 | |
| 9. two hundred twelve | 323 | |

## Thinking Cap

If you add one to the digit in the units place of 119 what happens? Why?

If you subtract 5 from the digit in the units place of 123 what happens? Why?

© 1993 CASIO, INC.   The Casio SL-450: A Tool for Teaching Mathematics

# TEACHER NOTES: Knowing Your Place

**Objective:** Demonstrate knowledge of the value of digits in specified places of given numbers.

**Grade Level:** 2-3

**Topic:** *Place Value*

## Using the Activity:

This activity can be completed with students working in pairs alternating use of the calculator and recording responses. Review with students the meaning of each place value for hundreds, tens, ones before starting the activity.

Students should know that since the 6 is in the tens place of 369 it represents 6 groups of ten. To remove the 6 you must subtract 60. To change the 3 in 309 to a 5, students must know that you need to add 2 more groups of 100, therefore, you enter + 200.

## Answers:

*What Did You Do?*

1. [−][2][0]
2. [−][3]
3. [−][3][0][0]
4. [−][6][0]
5. [−][9]
6. [+][4][0]
7. [+][3][0][0]
8. [+][2][0]
9. [+][1][1][1]

## Thinking Cap

These questions provide readiness experiences for regrouping. The calculator shows when one is added to 119, 120 is obtained. This helps explain that ten ones have been regrouped to form a group of ten.

When 5 is subtracted from 123, 118 is displayed. This helps explain that a group of ten has been regrouped to 10 ones.

## Extension:

Using only the [0], [1], [+], [=] keys, display 354 using the fewest key strokes.
*Answer: (without automatic constant)* 18 keystrokes  111 + 111 + 111 + 11 + 10 =
*(with automatic constant)* 14 keystrokes  111 + + = = + 11 + 10 =

**8**  The Casio SL-450: A Tool for Teaching Mathematics

# Read My Mind!

Name_____

The calculator will add ____ to every number entered.

| Enter | Predicted Display | Actual Display |
|---|---|---|
| 4 | | |
| 6 | | |
| 8 | | |
| 9 | | |
| 15 | | |
| 13 | | |
| 18 | | |
| 10 | | |

The calculator will subtract ____ from every number entered.

| Enter | Predicted Display | Actual Display |
|---|---|---|
| 9 | | |
| 13 | | |
| 17 | | |
| 15 | | |
| 18 | | |
| 8 | | |
| 14 | | |
| 10 | | |

## Thinking Cap

| Entered | What did the calculator do? | Display |
|---|---|---|
| 9 | | 5 |
| 16 | | 7 |
| 8 | | 17 |
| 15 | | 6 |

© 1993 CASIO, INC.  The Casio SL-450: A Tool for Teaching Mathematics

# TEACHER NOTES: Read My Mind

**Objective:** To practice basic facts through use of the automatic constant.

**Grade Level: K-3**

**Topic:** *Operation Readiness*

## Using the Activity:

This activity is intended to be directed by the teacher using an overhead OH-450 calculator. Students do not need to have the calculator in their hands. Students could have number cards to hold up indicating the answer or can record the answers on paper. The teacher sets the calculator to automatically add a given number to a set of numbers or subtract a given number from a set of numbers. Before the teacher presses equal to display the answer, students must predict what the calculator will display. The actual display is recorded to verify the prediction. The calculator can be reset to automatically add or subtract any number by clearing the calculator (AC) and reentering the sequence:

constant addend [+] [+]   number to be added to [=]
　　　　　　　　　　　　 number to be added to [=]
　　　　　　　　　　　　 number to be added to [=]

For example,   7 [+] [+] 5 [=]   12  (display)
　　　　　　　　　　　　6 [=]    13  (display)
　　　　　　　　　　　　9 [=]    16  (display)
　　　　　　[AC] 9 [+] [+] 4 [=]  13  (display)
　　　　　　　　　　　　6 [=]    15  (display)

**For subtraction enter, subtrahend [−] [−] minuend [=]**

For example,   7 [−] [−] 9 [=]    2  (display)
　　　　　　　　　　　　13 [=]    6  (display)

Students could be asked to write the corresponding problem in symbolic form, that is
**7 − − 9 = 2** would be written **9 − 7 = 2.**

## Thinking Cap

Students must indicate what the calculator did to the number entered to obtain the number displayed. Answers: subtracted 4; subtracted 9; added 9; subtracted 9

## Extension

To do the activity with multiplication or division facts enter the following sequences:

constant factor [×] [×]  number [=]   OR   divisor [÷] [÷]  number [=]
　　　　　　　　　　　　 number [=]　　　　　　　　　　　　 number [=]

# Squeeze Play

Name_____

For each problem find three different numbers which will cause the answer to fall in the given range.

1.  38 + △ = $\begin{array}{c} 100 \\ \hline 200 \end{array}$

    △ = ____ or △ = ____ or △ = ____

2.  63 + △ = $\begin{array}{c} 100 \\ \hline 150 \end{array}$

    △ = ____ or △ = ____ or △ = ____

3.  125 − △ = $\begin{array}{c} 50 \\ \hline 100 \end{array}$

    △ = ____ or △ = ____ or △ = ____

4.  △ − 53 = $\begin{array}{c} 50 \\ \hline 100 \end{array}$

    △ = ____ or △ = ____ or △ = ____

5.  △ + 29 = $\begin{array}{c} 50 \\ \hline 60 \end{array}$

    △ = ____ or △ = ____ or △ = ____

## Thinking Cap

Find two numbers whose sum is between 100 and 150.
Find two numbers whose difference is between 20 and 40.

© 1993 CASIO, INC.    The Casio SL-450: A Tool for Teaching Mathematics

**TEACHER NOTES:** *Squeeze Play*

**Objective:** To estimate sums and differences within a given range.

**Grade Level:** 2 - 3

**Topic:** *Estimation*

**Using the Activity:**

For each exercise students are asked to find three different values that can be added to or subtracted from a given number and have the answer fall within a specified range. Students could complete the activity using guess and check, randomly trying numbers to see if they satisfy the condition. Students who understand the inverse relationship of addition and subtraction could, for example in the first problem, subtract 38 from 100 and 38 from 200 to find the range of missing numbers. An additional strategy would be to have students apply estimation skills to determine the answer. For example in problem one, students should think of $38 + \Delta$ as approximately $40 + \Delta$ to facilitate finding a solution. Students should verify each answer on the calculator. Many answers are possible for each problem. The calculator frees the student from the computation, allowing the student to concentrate on the problem solving process.

Ask students to determine for each exercise the smallest and largest number that satisfies each condition. *Note: Sum falls in the range. The endpoints are not included.* (**1.** *63; 161*  **2.** *38; 86*  **3.** *26; 74*  **4.** *104; 152*  **5.** *22; 30*)

**Thinking Cap**

Students are asked to find two numbers whose sum is between 100 and 150. Many answers are possible such as 76 and 60. In the second question, they are asked to find two numbers whose difference is between 20 and 40. A possible answer is 55 and 80. Be sure to have students explain how they determined their pair of numbers. An additional challenge, not found on the student page, would be to ask students to find two numbers that satisfy two conditions at the same time. For example, find two numbers whose sum is between 50 and 60 and whose difference is between 5 and 10. A possible answer would be 30 and 24.

# Pitch It In!

Name_____

Find the total score if three balls are tossed into the given scores.

1.   

2.   

What must the score of the third ball be in order to get the given total score?

3.  38

4. 30

5. 39

6. 61

## Thinking Cap

How many ways can you score 36 if you have 3 balls to toss?

© 1993 CASIO, INC.   The Casio SL-450: A Tool for Teaching Mathematics  13

**TEACHER NOTES:** *Pitch It In!*

**Objective:** To use the calculator to solve problems involving the meaning of addition and subtraction.

**Grade Level:** 1-3

**Topic:** *Problem Solving/Whole Number Operations*

## Using the Activity:

This activity uses the calculator as a tool to solve problems involving the concept of addition and subtraction. In the first two problems, students are expected to find the total score that results from pitching 3 balls into the target. The object here is not to see if the children can add three whole numbers, but rather to see if they can reason that the situation requires addition to solve the problem. Once the students decide on the operation the focus shifts to determining if the student can correctly key the operation into the calculator.

In problems 3-6, the total score and two of the three ball scores are provided. Students must determine the third score. This requires students to recognize the use of subtraction to solve the problem. Students who do not understand this relationship, may select the guess and check strategy, randomly trying numbers from the target until they find the one that works. In discussing the activity, highlight the inverse relationship between addition and subtraction and how it applies to the given situation.

*Answers:*   **1.** *27*   **2.** *58*   **3.** *15*   **4.** *6*   **5.** *18*   **6.** *18*

## Thinking Cap

In this section, students are challenged to find combinations from the target board that total 36. The calculator is an effective problem-solving tool in this situation, allowing students to easily test combinations. The problem states that the child has 3 balls to throw. It doesn't preclude the possibility that one or more balls misses the target. The solutions are: 15, 15, 6; 25, 5, 6; 18, 18, miss

# Making Cents

Name_____

Use your calculator to determine the total amount of money in each problem.

1. 25¢ 10¢ 10¢ 5¢ 1¢ 1¢ 1¢

2. 25¢ 25¢ 10¢ 10¢ 10¢ 5¢ 5¢ 1¢ 1¢

3. 25¢ 25¢ 25¢ 5¢ 5¢ 5¢ 1¢ 1¢ 1¢ 1¢

4. 50¢ 25¢ 10¢ 5¢ 1¢

5. 10¢ 10¢ 10¢ 5¢ 5¢ 5¢ 5¢ 1¢ 1¢ 1¢ 1¢

6. 50¢ 25¢ 10¢ 10¢ 10¢ 5¢ 5¢ 5¢ 5¢ 1¢ 1¢

## Thinking Cap

Can you have 66¢ with 6 coins? 5 coins? 4 coins? 3 coins? Show how.

**TEACHER NOTES:** *Making Cents*

**Objective:** To determine the value of a collection of coins.

**Grade Level:** 1-3

**Topic:** *Problem Solving*

## Using the Activity:

This activity focuses on the use of the calculator to determine the value of a collection of coins. Students can be asked to determine the total value of the collection first without the calculator and then use the calculator to verify their answers. When entering the values of the coins, students can enter a quarter as either 25 or .25 depending on their understanding of the decimal notation for money. Students need to be consistent in how they enter the values. If one value is entered as a decimal, then all values must be in decimal form.

When entering money amounts in decimal form into the calculator, students need to understand that if .10 is entered, as soon as $=$ is pressed, the calculator will display the number as .1. If the decimal value of a set of coins ends in a zero, the calculator will eliminate ending zeroes to the right of the decimal point. This can be potentially confusing to students. It, however, provides an opportunity to introduce to students the ideas of equivalent decimals, that is .1 = .10 = .100.

*Answers:*
**1.** .53   **2.** .92   **3.** .94   **4.** .91   **5.** .54   **6.** 1.22

## Thinking Cap

Students are presented with a problem involving the need to express an amount of money in a specified number of coins. Students apply the guess-and-check strategy for problem solving to find the solution. The calculator is used to total coin values.

*Answer:*   6 coins: 2 quarters, 3 nickels, 1 penny
5 coins: 2 quarters, 1 dime, 1 nickel, 1 penny
4 coins: 1 half dollar, 1 dime, 1 nickel, 1 penny
3 coins: not possible

## Extension:

Use play money to act out the problems. Give students bags with collections of different coins in the bags. Have students determine the total value of each bag. Have them arrange the bags from least value to greatest value.

# Target Practice

Name_____

For each problem, select two numbers from the given set which you think will produce the given product. Multiply the numbers on your calculator. Record your selection as a hit, if you obtain the given product, or as a miss, if you do not obtain the given product. If you record a miss, try selecting two other numbers.

| Numbers | Product | PICK | HIT | MISS |
|---|---|---|---|---|
| 1. 34, 22, 32, 28, 44, 14 | 896 | | | |
| | | | | |
| 2. 23, 9, 36, 41, 18, 78 | 414 | | | |
| | | | | |
| 3. 63, 28, 9, 43, 18, 32 | 1024 | | | |
| | | | | |
| 4. 38, 41, 28, 51, 18, 61 | 2318 | | | |
| | | | | |
| 5. 61, 31, 57, 33, 62, 67 | 1881 | | | |
| | | | | |

## Thinking Cap

Find a whole number that when multiplied by 216 will produce a product between 14,000 and 14,500. What is the smallest whole number that multiplies 216 to produce a product in this range? What is the largest whole number that multiplies 216 to produce a product in this range?

© 1993 CASIO, INC.                The Casio SL-450: A Tool for Teaching Mathematics

# TEACHER NOTES: *Target Practice*

**Objective:** To estimate whole number products.

**Grade Level:** 4-5

**Topic:** *Estimation/Whole Number Operations*

## Using the Activity:

This activity focuses on number sense and estimation. Students are given sets of six factors and designated products. Students are to select the two numbers they feel multiply to give the target product. While students could complete the activity using guess and check, the thrust of the activity is to encourage students to use estimation and number sense to minimize guesses. Rounding the factors and the product narrows the selection combinations. Furthermore, the possibilities can be limited by recognizing that the units digit of the product is equal to the unit digit of the product of the units digits of the two factors. That is in question 1, the designated product is approximately 900. The factors which might be selected are 34 x 28, or 32 x 28, or 22 x 44. But, the units digit of the product is 6. Hence, 32 x 28 gives a units digit of 6. Therefore, 28 and 32 are the best picks.

Students should be expected to verbalize the strategy they used to select the factors. The calculator removes the computational demand from the problem and allows the student to focus on the estimation and number sense.

*Answers:*    **1.** *32, 28*    **2.** *23, 18*    **3.** *28, 48*    **4.** *38, 61*    **5.** *33, 57*

## Thinking Cap

In this section, students are asked to provide the missing factor that will produce a product within a given range. Students can use a variety of approaches including guess and check, rounding factors and products, or dividing the product by the given factor to find the missing factor. The smallest whole number is 65. The largest whole number is 67.

# Leftover Crunchies    Name_____

John wanted to use his calculator to solve the following problem.

**The Crunchy Cookie Company packages 24 cookies to a box. On Tuesday, 900 cookies were baked. How many boxes of cookies could be packaged? How many cookies remained unpackaged?**

What should John enter into the calculator to find the answer?_____

What does the calculator display as the answer?_____

How many boxes of cookies were packaged?_____

How many of the 900 cookies were packaged?_____

How many of the 900 cookies are still unpackaged?_____

How many more cookies must be baked to complete the remaining box?_____

Use your calculator to solve the following problems.

**1.** At an egg packing plant, eggs are boxed in dozens. How many dozen eggs can be boxed if the plant has 400 eggs in stock?_____

**2.** A school with 895 students enrolled is planning a field trip for all the students enrolled in the school. The average school bus seats 42 students. How many buses are needed for the trip?_____

**3.** A stamp album holds 12 stamps per page. Mary Ann has 256 stamps to enter into the album. How many stamps will be entered on the last page containing stamps?_____

**4.** A punchbowl holds 6 gallons of punch. Kelly plans to use 15 ounce containers to store the punch. How many 15 ounce containers will she need?_____
How much more punch is needed to fill the last container?_____

## Thinking Cap

In a division problem, if the dividend is 12548 and the quotient is 522.83333, what is the divisor and the remainder?_____

# TEACHER NOTES: *Leftover Crunchies*

**Objective:** To solve problems involving interpretation of remainders.

**Grade Level:** 4-5

**Topic:** *Whole Number Operations/Problem Solving*

## Using the Activity:

The calculator displays remainders to whole number division problems as decimal fractions. This activity is designed to teach students how to retrieve a whole number remainder from the display given. Students should work through the Crunchy Cookie problem using the calculator. Discuss the answers to the questions before having the students complete the other four problems on the sheet.

*Answers:*
What should John enter into the calculator to find the answer? $900 \div 24$
What does the calculator display? 37.5
How many boxes of cookies were packaged? 37
How many of the 900 cookies were packaged? $37 \times 24 = 888$
How many of the 900 cookies are still unpackaged? $900 - 888 = 12$
How many more cookies must be baked to complete the remaining box? $24 - 12 = 12$

Summarize how to obtain the whole number remainder from the calculator display.

**First, [ ( whole number part of quotient) x divisor]**
**Second, subtract answer from dividend to get remainder.**
**$900 \div 24 = 37.5$;   [(37 x 24)] = 888;   900 - 888 = 12 remainder**

For the four problems, have students state what they entered into the calculator and what the calculator displayed in addition to giving the answer to the problem.

*Answers:*   **1.** 33 dozen   **2.** 22 buses   **3.** 4 stamps (256-(12 x 21))   **4.** 52 containers; 12 ounces  (15 - ( 768 - (15 x 51)))

## Thinking Cap

Students need to divide the dividend by the quotient to obtain the divisor and then subtract the product of the divisor and the whole number part of the quotient from the dividend to obtain the remainder.  *Answer: divisor = 24, remainder = 20*

## Extension

Have students complete $300 \div 24$ using both the calculator and paper and pencil. The calculator answer is 12.5, while the paper answer is $12\frac{1}{2}$. Ask students what the .5 in the calculator display represents. *(remainder)* What is that equivalent to in the paper and pencil computation? *($\frac{12}{24}$ or $\frac{1}{2}$)*  Discuss equivalence relationship.

# Point The Way!

Name_____

For each problem, place decimal points in the addends as needed to obtain the indicated sum. Use your calculator to check your answer.

1.  245 + 82 = 3.27

2.  519 + 283 = 33.49

3.  893 + 247 = 114

4.  5 + 7 + 9 = 2.1

5.  26 + 357 + 4 = 38.7

For each problem, place a decimal point in the difference to provide a reasonable answer. Use your calculator to check your answer.

6.  2.4 − 1.36 = 104

7.  5 − 3.742 = 1258

8.  5.01 − 3.89 = 112

## Thinking Cap

Place + and/or − signs between the numbers to obtain the given answer.

8.4 ? 3.7 ? 2.9 ? 8.4 ? 6.7 = 9.3

# TEACHER NOTES: Point The Way!

**Objective:** To determine reasonable sums and differences of decimals.

**Grade Level:** 4-5

**Topic:** *Decimals/Number Sense*

## Using the Activity:

This activity is designed to have students estimate where decimal points should be placed in addition and subtraction problems to provide reasonable answers to specified conditions. As students use the calculator to check their answers, they gain additional facility using the calculator.

In problems 1-5, students are to place decimal points in the addends as needed to produce the indicated sum. A strong sense of place value will help students determine the relative magnitude of the addends needed for the given sum. Discussion should focus on the strategy used to determine where the decimal point(s) should be placed.

In problems 6-8, the decimal point is missing from each difference. It is not the intent of the activity that students do actual computation to determine where to place the decimal point, but rather that they use reasoning and estimation skills to determine the placement of the decimal point.

In problem 4, the answer is .5 + .7 + .9. Students should see that when they enter .5 into the calculator, the display reads 0.5. The calculator will always maintain a zero in the units place of the number, if the number has no whole number part. Furthermore, in problem 3, where the sum is 114.0, the calculator drops the zero in the tenths place. Students should understand that when dealing with decimals, ending zeros to the right of the decimal point are not necessary, hence, the calculator will drop them. This is a good time to discuss with students equivalent decimals.

*Answers:*   **1.** 2.45 + .82   **2.** 5.19 + 28.3   **3.** 89.3 + 24.7   **4.** .5 + .7 + .9
**5.** 2.6 + 35.7 + .4   **6.** 1.04   **7.** 1.258   **8.** 1.12

## Thinking Cap

In this section, students are to apply problem solving skills to determine the combination of addition and subtraction signs that need to placed between the given numbers to obtain the indicated answer. The calculator allows students to easily test and modify answers.

*Answer:* 8.4 − 3.7 + 2.9 + 8.4 − 6.7 = 9.3

# What's The Point?   Name_____

Enter each exercise into the calculator. Record the products in the displays.

1. 237 x 15 =
2. 237 x 1.5 =
3. 237 x .15 =
4. 2.37 x 1.5 =
5. 2.37 x .15 =
6. .237 x .15 =

Study the displays. Compare the results.

7. How are the displays the same?_____

8. How are the displays different?_____

9. When you multiply decimals, what is the rule?_____
_____

10. According to your rule, where would you place the decimal point in the product 35.6 x .24 = 8544?_____

## Thinking Cap

Circle the reasonable answer. Check the result using your calculator.

62.5 x 3.76 =   .235   2.35   23.5   235

Bryan was confused when he used his calculator to find 3.12 x 2.5. The calculator displayed the product to be 7.8. He was expecting a product with three decimal places. Can you explain what happened?_____
_____

How do you know that 7.8 is a reasonable answer for the product of 3.12 and 2.5?
_____

© 1993 CASIO, INC.    **The Casio SL-450: A Tool for Teaching Mathematics**

**TEACHER NOTES:** *What's The Point?*

**Objective:** Find the product with decimal factors.

**Grade Level:** 5-6

**Topic:** *Decimals*

### Using the Activity:

This activity is designed to have students discover the rule for multiplying with decimal factors. Students are expected to use number sense and reasoning to complete the activity.

Have students complete the first set of problems using the calculator.

Note: When the student enters .15 the calculator displays 0.15. The calculator maintains a zero in the units place of decimals.

Have them compare results and answer the questions.

How are the displays the same? *The digits in each pair of factors and the digits in the products are the same and in the same order.*

How are the displays different? *The decimal points in the products are in different locations. The value of the products and factors is different.*

What is the rule? *Multiply the factors as you multiply whole number factors. Place the decimal point in the product the number of places equal to the sum of the decimal places in the factors.*

### Thinking Cap

This section focuses on the need to be sure the product is reasonable. When the calculator multiplies decimals, the display does not show zeros after the decimal point. Hence when students look at the display, they may be confused as to why there are fewer numbers after the decimal than they expected. Students ought to know that since 3.12 is approximately 3 and 2.5 is approximately 3, the answer must be close to 9.

### Extension:

Use the digits **9, 8, 7, 6, 5** once and only once to complete the problem producing the largest possible product.

$$\begin{array}{r} \triangle\,.\,\triangle\,\triangle \\ \times\ \triangle\,.\,\triangle \\ \hline \end{array}$$

(8.75 x 9.6 = 84)

# Don't Touch That Key!   Name_____

Suppose the [7] on your calculator is broken. You cannot use the key to enter the digit 7. The calculator will display a 7 if it results from a computation. You are just not able to enter it directly.

Indicate what you would enter into the calculator to produce the given display. In each case, show 2 different ways of obtaining the display.

| Display | Enter |
|---|---|
| 1.  707 | |
|    707 | |
| 2. the answer to 47 + 29 | |
|    the answer to 47 + 29 | |
| 3. the answer to 376 + 457 | |
|    the answer to 376 + 457 | |
| 4. the answer to 767 − 589 | |
|    the answer to 767 − 589 | |
| 5. the answer to 854 − 278 | |
|    the answer to 854 − 278 | |
| 6. the answer to 27 × 64 | |
|    the answer to 27 × 64 | |

## Thinking Cap

Suppose the only keys on your calculator that worked were:

, , , , , [×], [÷], [=]

How could you find 36 × 63?

**TEACHER NOTES:** *Don't Touch That Key!*

**Objective:** To apply mathematical properties to the evaluation of expressions.

**Grade Level:** 4-6

**Topic:** *Numeration/Problem Solving*

## Using the Activity:

This activity has students focus on the application of mathematical properties and mental-math strategies to evaluate expressions by restricting access to specified keys on the calculator. The computation involved in the given expressions is easy enough to do without the calculator. However, actual computation is not the focus. The main idea is to help students understand why it is possible to represent the problems in different ways. In this activity, students might apply the associative properties of addition or multiplication, the distributive property, and/or mental-math strategies such as compensation. Being given opportunities to recognize when to use these properties empowers the students with mathematical understanding.

As with all activities, it is important to have students discuss their thought processes and continually seek alternate solutions to the problems. This activity can be completed with students working in small groups.

*Answers (can vary)*
**1.** $699 + 8$  **2.** $(46 + 1) + 29$, $48 + 28$, $46 + 30$  **3.** $(350 + 26) + (455 + 2)$, $380 + 453$
**4.** $666 + 101 - 589$, $800 - 622$  **5.** $854 - 268 - 10$, $854 - 280 + 2$
**6.** $9 \times 3 \times 64$, $26 \times 64 + 64$, $(25 + 2) \times 64$

## Thinking Cap

To complete this activity, the students may find a variety of key combinations that work. One possible solution involves thinking about the problem in a different way, based on the associative and commutative properties for multiplication, if students think of $36 \times 63$ as $9 \times 4 \times 9 \times 7$ and commute factors, they get $9 \times 9 \times 4 \times 7$ or $81 \times 28$. Hence, $36 \times 63$ can be found by entering $81 \times 28$.

# Can You Top This?   Name_____

1. Select 6 different digits from the given set. Using the 6 digits form two 3-digit numbers which produce the largest possible sum.

    [0, 1, 2, 3, 4, 5, 6, 7, 8, 9]

    ```
      △ △ △
    + △ △ △
    ─────────
    ```

2. Is this the only possible solution?_____ Why or why not? _____
   _____

3. Using the four given digits, form two 2-digit numbers with the smallest possible difference.

    [3, 5, 7, 9]

    ```
      △ △
    − △ △
    ─────
    ```

4. How would you rearrange the digits to obtain the largest possible difference?
   _____

5. Using the four given digits once in each problem, complete the multiplication problems given to form the largest possible products.

    [3, 5, 7, 9]

    ```
      △ △ △              △ △
    ×     △            × △ △
    ─────────         ─────────
    ```

## Thinking Cap

Use the digits 1, 3, 5, 7, 9 to form two 2-digit numbers and one 1-digit number such that the final product is the largest possible *even* product.

   △ × (△ △ − △ △)

© 1993 CASIO, INC.              The Casio SL-450: A Tool for Teaching Mathematics

**TEACHER NOTES:** **Can You Top This?**

**Objective:** To apply number and operation sense to given problems.

**Grade Level: 4-5**

**Topic:** *Whole Number Operations/Problem Solving*

### Using the Activity:

The focus of this activity is to use the calculator for rapid guess and check in order to find combinations of numbers that satisfy specified conditions. Students should record all combinations tried. Students may be inclined to stop working on the problem when they reach a solution, not knowing if it is the one that actually satisfies the condition. Have students work in groups. After a reasonable amount of time, call for answers from the groups.

When a correct answer is reached, it is important to analyze with students the relationships which help convince them that this is in fact the correct solution. Repeat the activity using a different set of digits for each problem. This gives students the opportunity to reinforce the concepts discussed.

*Answers:* **1.** *975 + 864 = 1839* **2.** *Not the only possible answer. The key is that the largest digits must be in the largest places in any combination. For example, 875 + 964 = 1839 or 965 + 874 = 1839 or 974 + 865 = 1839.* **3.** *93 − 75 = 18* **4.** *97 − 35 = 62* **5.** *753 x 9 = 6777, 75 x 93 = 6975*

### Thinking Cap

In this section, students must realize that to obtain an even product with one factor odd requires the other factor, obtained from the difference, to be even. Furthermore, to obtain the largest product requires the students to form the largest even difference possible.   7 x (95 − 13) = 574

# The Race Is On!

Name_____

Work with a partner. One of you does each problem in column A on the calculator, keying in every digit and operation indicated. The other person must find the answer to the problem in column A without using the calculator. Look for ways to make your computation easier. See who completes the problems first. Switch roles and complete column B. What did you discover?

| Column A | Column B |
|---|---|
| 1. 450 + 250 | 1. 360 + 240 |
| 2. 978 − 378 | 2. 8734 − 3734 |
| 3. 862 − 751 | 3. 619 − 508 |
| 4. 263 + 975 − 263 | 4. 983 − 458 + 458 |
| 5. 3 x 2 x 4 x 2 | 5. 2 x 5 x 3 x 4 |
| 6. 10 x 5 x 2 x 63 | 6. 20 x 5 x 18 |
| 7. 6754 x 385 x 0 | 7. 6374 x 273 x 0 |
| 8. 300 x 40 | 8. 800 x 700 |
| 9. 5 x 95 x 2 | 9. 50 x 2 x 6 |
| 10. 7600 ÷ 100 | 10. 630 ÷ 10 |

## Thinking Cap

Suppose you did not have a calculator and you had to find the following product, 239 x 99. How could you think about the problem in a different way to make the computation easier to do?

© 1993 CASIO, INC.

**The Casio SL-450: A Tool for Teaching Mathematics**

# TEACHER NOTES: *The Race Is On!*

**Objective:** To compute answers using the calculator and mental math.

**Grade Level:** 4-6

**Topic:** *Problem Solving*

## Using the Activity:

This activity is designed to help students see that using the calculator is not always the quickest way to obtain an answer. It is critical for students to develop a sense as to when to use mental math versus when to use the calculator. The students' objective is usually to finish an assignment as quickly as they can, hence they often don't want to spend the time looking at the problem carefully before solving it. Problem number 7 shows that if you look at the entire problem first, you can immediately see that the product is 0 and there is no need to do any multiplication. Students need to recognize that in mathematics, "you have to spend time to save time."

Students should not be given carte blanche for use of the calculator. They should be held accountable to make responsible decisions as to when it is appropriate and necessary to use the calculator, and when either mental math or paper-pencil computation can give them the answer more quickly. Be sure to discuss these points with students.

When completing this activity with students, pair students of equal computational proficiency together. This is to prevent giving any one student an unfair advantage. Students should see that for the problems in the activity, little or no time is saved in using the calculator.

## Thinking Cap

The problem presented here asks students to use mental-math strategies to make the problem easier to compute or capable of being computed in a different way. One possible response would be 239 x 100 – 239.

## Extension

Present students with a sheet of 10 problems, some that are easily done by mental math and some that lend themselves to use of the calculator. Ask students to vote for which method they would use and discuss why they selected the method.

# First Class Mail

Name_____

To mail a letter first class, the post office computes the cost based on the weight of the letter. A letter weighing one ounce or less required 29¢ postage in 1993. For each additional ounce or part of an ounce an additional 23¢ is charged.

Using the following sequence on your calculator you can compute the costs of mailing letters weighing more than one ounce.

Enter:  .23  [+]  [+]  .29  [=]  [=]  . . .

Complete the chart.

| Weight of Letter/Package | Cost of First Class Postage |
|---|---|
| up to 1 oz | 29¢ |
| more than 1 oz to 2 oz | |
| more than 2 oz to 3 oz | |
| more than 3 oz to 4 oz | |
| more than 4 oz to 5 oz | |
| more than 5 oz to 6 oz | |
| more than 6 oz to 7 oz | |
| more than 7 oz to 8 oz | |

1. Suppose a package costs $1.90 to mail. What could its weight be?

   _____

2. Suppose a package costs $3.51 to mail. What could its weight be?

   _____

3. Which of the following amounts couldn't be charged for first class postage on a letter?  $2.59,  $3.05,  $3.94,  $4.20

## Thinking Cap

Kevin mailed two letters for a total cost of $4.72. Find all possible cost combinations of the two letters mailed.

# TEACHER NOTES: First Class Mail

**Objective:** To use the calculator to generate data needed to solve problems.

**Grade Level:** 4-6

**Topic:** *Problem Solving*

## Using the Activity:

This activity presents the calculator as a tool in problem solving. Students use the automatic constant for addition to generate data on postage rates needed to answer given problems. The cost of first class postage is computed by adding an additional 23¢ to the base cost of 29¢ for each additional ounce or part of an ounce over one that the letter weighs. It is a natural application of the automatic constant function on the calculator.

Problem number one can be answered from the data generated in the chart. However, to answer problem 2, students would have to extend the chart by using either patterns or the calculator, or would have to employ the problem-solving strategy of working backwards. In working backwards, the students would subtract 29¢ from $3.51 and then divide the difference by .23 to determine the number of times 23¢ has been added to the base cost. In this problem, 23¢ has been added 14 times. It follows, therefore, that the letter weighs more than 14 oz. up to a possible 15 oz. Problem 3 requires students either to recognize the pattern to determine which numbers would occur in the chart, or to subtract 29¢ from each number and see if the difference is evenly divisible by 23¢.

*Answers: Chart; 52¢, 75¢, 98¢, $1.21, $1.44, $1.67, $1.90*
*1. more than 7 oz to 8 oz   2. more than 14 oz to 15 oz   3. $3.94*

## Thinking Cap

In this section, the postage rate function is explored. Since the weight of each package is computed as 29¢ plus n times 23¢, where n represents the additional ounces over one that the letter weighs, there are many combinations that give a total of $4.72.

*Answers: 29¢ and $4.43,  52¢ and $4.20,  75¢ and $3.97,  98¢ and $3.74,  $1.21 and $3.51, $1.44 and $3.28,  $1.67 and $3.05,  $1.90 and $2.82,  $2.13 and $2.59,  $2.36 and $2.36.*

## Extension

Have students graph the relationship placing weight on the horizontal axis and cost on the vertical axis. The result shows the graph is a step function.

# Mean Machine

Name_____

Average or mean = $\dfrac{\text{sum of set of numbers}}{\text{number of numbers in the set}}$

Use your calculator to find the average of the following:

1. 80  82  88  98           AVERAGE:_____

   a. Which of the given numbers are above average?_____

   b. Altogether, how much are they above average?_____

   c. Which of the given numbers are below average?_____

   d. Altogether, how much are they below average?_____

   e. Are any of the numbers in the set equal to the average?_____

   f. Comparing the amount above the average with the amount below the average, what do you observe?_____

   g. Suppose 5 is added to each number in the set. What will happen to the average for the set?_____

   h. Give a set of 4 other numbers that has the same average as the given set of numbers?_____

2. 420  436  451  398           Average:_____

3. 69.3  82.1  71.4  93.6         Average:_____

4. 69.3  70  70.7  75  64         Average:_____

## Thinking Cap

Jacqueline has taken 4 tests in math this quarter. Her scores are 82, 96, 84, 90. There is one more test in the quarter and Jacqueline wants to have an average of 90 for the quarter. What must she get on the fifth test?

# TEACHER NOTES: Mean Machine

**Objective:** To find the average of a set of numbers using the calculator.

**Grade Level:** 4-6

**Topic:** *Problem Solving*

## Using the Activity:

This activity uses the calculator as a tool to find the average of a set of numbers. Students are given the definition of average at the beginning of the activity. By using the calculator to find the averages, students can focus on the concept of average and the application of average rather than on the computation. A set of questions asked in connection with the average of the first set of numbers, gives students the opportunity to uncover important relationships and concepts about average. The average for the problem is 87. Hence, 88 and 98 are above average. Together, the two numbers are 12 above average (88 – 87 = 1 and 98 – 87 = 11). Conversely, 80 and 82 are below average. Together, the two numbers are 12 below average ( 87 – 80 = 7 and 87 – 82 = 5). This relationship helps students to see that the average is the equalizer of the differences of the numbers above and below average with the average. Students explore the concept further to learn that if the same number is added to each number in the set, the average is increased by that number. The final question allows students to use trial and error to determine a set of numbers that has an average of 87.

Answers:   **1.** average 87   **1a.** 88, 89   **1b.** 1 and 11 total 12   **1c.** 80, 82   **1d.** 7 and 5 total 12   **1e.** no   **1f.** total difference of numbers above and below average are the same   **1g.** average is increased by 5   **1h.** answers will vary; 84, 87, 87, 90
**2.** 426.25   **3.** 79.1   **4.** 69.8

## Thinking Cap

This section presents the reverse of the problems 1-4. If you know the average of 5 scores and the value of 4 of the 5 scores, to find the fifth score students should add the four known scores (82 + 96 + 84 + 90 = 352) and store that in memory. The desired average should be multiplied by the total number of scores (90 x 5). The sum is recalled from the memory and subtracted from the product to give the value of the fifth score (450 – 352 = 98)  Answer: 98.

# Name That Part

Name_____

Every fraction can be named by a decimal. When your calculator works with fractions, it uses the decimal name for the fraction. To get the decimal name for a fraction **a/b,**

**Enter: a ÷ b**

For example, to find the decimal name for 1/2, enter 1 ÷ 2 to obtain 0.5.

**Use your calculator to find the decimal name for each fraction.**

| | FRACTION | DECIMAL |
|---|---|---|
| 1. | 1/4 | |
| 2. | 3/4 | |
| 3. | 1/5 | |
| 4. | 4/5 | |
| 5. | 1/8 | |
| 6. | 3/8 | |
| 7. | 6/8 | |
| 8. | 7/8 | |
| 9. | 7/16 | |

**Using the decimal names for each fraction, compare the values and circle the larger number.**

10.  3/8   7/8        11.   7/8   7/16        12.   3/4   6/8

_____  _____              _____  _____              _____  _____

## Thinking Cap

Use the decimal names for the following fractions to put them in order from least to greatest.   7/8, 3/4, 4/5, 7/16

TEACHER NOTES: **Name That Part**

**Objective:** To use the decimal equivalent of given fractions to compare fractions.

**Grade Level:** 4-6

**Topic:** *Fractions*

### Using the Activity:

The SL-450 converts any fraction input into the calculator to a decimal. All fractional computations on the calculator are completed as decimals and would require students to convert the decimal back to a fraction with paper and pencil if they desired the final answer to be in fraction form.

This activity has students use the decimal equivalent for a given fraction to compare fractions. Students are instructed to input the fraction **a/b** as **a ÷ b** and record the resulting decimal equivalent in a chart. Using the data in the chart, students are asked to compare three pairs of fractions. In one pair, the fractions have the same denominator and hence the fraction with the larger numerator is larger as evidenced by its decimal form. In the second pair of number 11, the fractions have the same numerators, and the decimal comparison supports the fact that with the same numerator the smaller the denominator the larger the fraction. Number 12 shows that equivalent fractions have the same decimal form. Using the fractions given in the chart, make up other pairs of fractions for students to compare.

Have students analyze the relationship between the decimals for 1/5 and 4/5. Students should see that the decimal form for 4/5 is 4 times the decimal form for 1/5. Using that relationship, ask students what the decimal form for 3/5 and 7/5 would be.

*Answer:*  **1.** 0.25   **2.** 0.75   **3.** 0.2   **4.** 0.8   **5.** 0.125   **6.** 0.375   **7.** 0.75   **8.** 0.875   **9.** 0.4375   **10.** 7/8   **11.** 7/8   **12.** equal

### Thinking Cap

In this section students are asked to use the data in the chart to order four fractions from least to greatest. Since the decimals in order from least to greatest would be: 0.4375, 0.75, 0.8, 0.875, it follows that the order of the fractions must be 7/16, 3/4, 4/5, 7/8.

### Extension

Have students represent each fraction using a manipulative such as fraction strips. Compare the lengths of the strips representing the fractions to verify the comparisons made using the decimal equivalents.

# Expand Your Memory! Name_____

To have the calculator write a number given in expanded notation

(3 x 1000) + (4 x 100) + (9 x 10) + (3 x 1)

in standard notation, 3493, **enter:**

> 3 [x] 1000 [M+]
> 4 [x] 100 [M+]
> 9 [x] 10 [M+]
> 3 [x] 1 [M+]
> [MR]

The display reads 3493.

**Press [AC] to clear the memory.**

Use your calculator and the key sequence given above to find the standard notation for the following numbers. Be sure to press [AC] between each problem to clear the memory.

1. (6 x 1000) + (5 x 100) + (2 x 10) + (4 x 1)  _____

2. (5 x 10000) + (3 x 100) + (4 x 1)  _____

3. (3 x 1000) + (6 x 100) + (9 x 10)  _____

4. (7 x 10000) + (6 x 100) + (3 x 10) + (2 x 1)  _____

5. (3 x 100) + (5 x 1000) + (6 x 1) + (8 x 10)  _____

6. (5 x 10) + (8 x 10000) + (3 x 100) + (4 x 1)  _____

## Thinking Cap

John entered 3 [x] 100 [M+] 4 [x] 10 [M+] 5 [x] 1 [M+] [MR]. He forgot to press [AC] before entering 2 [x] 100 [M+] 1 [x] 10 [M+]. What will John get as a display when he presses [MR]? Explain why.

# TEACHER NOTES: *Expand Your Memory!*

**Objective:** To use the memory keys to express numbers given in expanded notation in standard notation.

**Grade Level:** 3-6

**Topic:** *Problem Solving/Place Value*

## Using the Activity:

In this activity, students use the memory on the calculator to express numbers given in expanded notation in standard notation. The SL-450 does not possess algebraic logic. Hence, if the sequence was keyed directly into the calculator as written, the calculator would first multiply 3 times 1000 and then immediately add 4 to that and then multiply 3004 times 100 and so forth. By using the memory, the multiplication can be done before the addition by having the products computed in the display area and stored in the memory. However, the calculator has only one slot for memory. Therefore, if the product 3000 is stored in the memory, the calculator will literally add the next value entered into the memory to the 3000. To check the value stored in the memory at any time, you press [MR]. As long as a number is stored in memory, a letter M will appear in the upper left side of the display screen. Notice in the given example, while 4 separate products are stored in the calculator, when the memory is finally recalled the display represents the sum of the 4 products, not any individual product. It is important that students remember to press [AC] to clear the memory in between problems or else the calculator will continue to add the numbers to the values currently in memory.

In this activity, students are exploring the following three keys:

[M+] adds values to the calculator memory

[MR] recalls and displays the value stored in the calculator memory

[AC] clears the memory

Note: It would be appropriate for grades 5 and 6 to include numbers with decimal places in this activity.

*Answers:* **1.** 6524  **2.** 50304  **3.** 3690  **4.** 70632  **5.** 5386  **6.** 80354

## Thinking Cap

This section illustrates what happens if the memory is not cleared between problems. The calculator adds the product 200 and the product 10 to the number 345 already stored from the first sequence. When [MR] is pressed 555 is displayed.

# Key Purchases

Name_____

## SCHOOL STORE PRICE LIST

| | | | |
|---|---|---|---|
| PENS | 89¢ | BOOKCOVERS | 10¢ |
| PENCILS | 15¢ | MARKERS | 79¢ |
| PADS | 35¢ | PROTRACTORS | 45¢ |
| ERASERS | 10¢ | PENCIL SHARPENER | 20¢ |
| RULERS | 15¢ | COMPASS | $1.00 |

John works in the school supply store. He uses his calculator to find the total cost of purchases. He uses the memory (M+) to find the total cost and uses the display area to find the cost for multiple purchases of the same item. The store is running a special involving coupons. John knows that when a coupon is presented, the amount of the coupon is subtracted from the purchase. John subtracts the coupon amount from the total cost of the purchase in the memory using M–.

Sarah purchased 2 pens, 3 pencils, and 1 pad. She gave John a coupon that read "25¢ off the purchase of 2 pens". John entered the following into his calculator to find the total cost.

2 [x] .89 [M+] 3 [x] .15 [M+] 1 [x] .35 [M+] .25 [M–] [MR]

The calculator display read 2.33.

**Use the calculator's memory to find the total cost of the given purchases. Be sure to clear the memory by pressing AC between each problem.**

1. Purchase: 3 pads, 4 bookcovers, 1 ruler, 1 compass
   Coupons: 15¢ off purchase of a compass
   Total Cost: _____

2. Purchase: 4 markers, 1 protractor, 3 pencils
   Coupons: 5¢ off the purchase of each marker
   Total Cost: _____

3. Purchase: 6 pads, 7 markers, 2 pencil sharpeners
   Coupons: 25¢ off the purchase of 6 or more markers
   Total Cost: _____

## Thinking Cap

Sandy purchased 4 pads, 6 pencils, 5 pens, and 3 markers. She had a coupon for 50¢ off a purchase of $5.00 or more. What is her change from $20? Explain.

TEACHER NOTES: **Key Purchases**

**Objective:** To use M+, M–, MR keys to compute the cost of specified purchases.

**Grade Level:** 4-6

**Topic:** *Problem Solving*

**Using the Activity:**

In this activity students work with the three memory keys to compute the cost of given purchases. To understand the memory on the calculator, students must realize that the calculator's memory is made up of only one slot. This means that if a number is stored in the memory and another number is also placed in the memory, the calculator will combine the two numbers in the slot. If [M+] is pressed, the two numbers will be added together and stored in the memory. If [M–] is pressed, the second number will be subtracted from the number already in the memory and the difference will be stored in the memory. To check the value stored in the memory, you press [MR]. This recalls and displays the number stored in the memory. As long as a value is stored in the memory the display screen will show a little M. To clear the memory, which should always be done when a new problem is begun, you press [AC].

The example illustrates the procedure used to directly find the final cost of a purchase involving the use of a discount coupon. The cost of the multiple purchases of the same item is computed in the display area and stored in the memory. As each additional item purchased is placed in the memory, the sum for the purchase is being computed. The coupon value is subtracted from the total in the memory through the use of [M–]. When [MR] is pressed, the total cost less the coupon value is given.

Have students work in pairs finding the total cost for each of the three given transactions. Have them follow the example given. Be sure they clear the memory before beginning a new problem. The total cost could be found without using the memory by computing each multiple purchase on the calculator and recording the answer on paper. The products could then be reentered and added. The coupon could be subtracted, leaving the cost. This, however, does not allow students to gain facility with the memory functions on the calculator nor does it allow them to efficiently use the calculator in a multiple-step, problem-solving situation.

*Answers:*  **1.** *2.45*  **2.** *3.86*  **3.** *7.78*

**Thinking Cap**

Key sequence used: 4 [X] .35 [M+] 6 [X] .15 [M+] 5 [X] .89 [M+] 3 [X] .79 [M+] [MR] (check if purchase ≥ $5) .50 [M–] 20 [–] [MR] [=]  *11.38 answer*

**The Casio SL-450: A Tool for Teaching Mathematics**

# Putting Things In Order

Name_____

Bryant knows that the correct answer to the following problem is 21.

**9 + 6 x 2**

But when he entered the problem into the calculator, the calculator displayed 30. Explain why._____

_____

Help Bryant to figure out the correct way to enter the following problems into the calculator to obtain the given correct responses. Check your answers on your calculator.

| Problem | Enter | Answer |
|---|---|---|
| 1. 27 ÷ 9 x 4 | | 12 |
| 2. 18 x 5 − 41 | | 49 |
| 3. 35 + 225 ÷ 15 | | 50 |
| 4. (23 + 67) x 9 | | 810 |
| 5. ( 387 − 378)$^3$ ÷ 81 | | 9 |
| 6. 360 ÷ 15 + 13 x 14 | | 206 |
| 7. 32 x 9 − 24 x 8 | | ? |
| 8. 4096 ÷ (352 ÷ 44)$^3$ | | ? |

## Thinking Cap

Using all the numbers 12, 15, 16, and 18, once and only once, and any combination of +, −, x, ÷ , and parentheses, write an expression whose value is 24.

The Casio SL-450: A Tool for Teaching Mathematics

# TEACHER NOTES: *Putting Things in Order*

**Objective:** To evaluate expressions according to operational order.

**Grade Level:** *5-6*

**Topic:** *Order of Operations/Problem Solving*

## Using the Activity:

The SL-450 does not possess algebraic logic. Hence, when evaluating expressions with mixed operations, students must recognize they cannot always enter the expression into the calculator as it is written. According to order of operations, parentheses, powers, and roots are evaluated first in order from left to right. Then multiplications and/or divisions are done in order from left to right. Finally, additions and subtractions are completed in order from left to right. Students must analyze the expression to determine what gets evaluated first and then input the numbers into the calculator.

In this activity, students are given an expression and the correct answer for the simplification of the expression using operational order. Students are asked to indicate how the expression should be entered into the calculator. For the last two expressions, the answers are not provided. Students must determine the answers themselves. Students should be encouraged to use the problem-solving strategy of guess and check and working backwards in an attempt to determine the correct entry procedure if they have not mastered the rules of operational order. To complete this activity, students need to understand how to use the memory and automatic constant for exponents.

*Answers*

Introductory Problem: The calculator first added 9 + 6 and then multiplied by 2.

**1.** 27 [÷] 9 [×] 4 [=]   **2.** 18 [×] 5 [−] 41 [=]   **3.** 225 [÷] 15 [+] 35 [=]   **4.** 23 [+] 67 [×] 9 [=]
**5.** 387 [−] 378 [×] [×] [=] [=] [÷] 81 [=]   **6.** 360 [÷] 15 [M+] 13 [×] 14 [M+] [MR]
**7.** 32 [×] 9 [M+] 24 [×] 8 [M−] [MR]   **8.** 352 [÷] 44 [×] [×] [=] [M+] 4096 [÷] [MR] [=] 8

## Thinking Cap

In this section, students are asked to write an original expression using all the numbers 12, 15, 16, and 18, once and only once, with any combination of +, -, ×, ÷, and parentheses that has value 24 under operational order. Students are expected to use problem solving strategies to determine the expression.

*Answer:* 15 × 16 − 18 × 12

# Power Up!

Name_____

Enter the numbers into the calculator as given. Record the display. Provide the missing exponent for the given base.

| Enter | Display | Power Indicated |
|---|---|---|
| 2 [x] [x] 1 [=] | | $2^\triangle$ |
| [=] | | $2^\triangle$ |
| [=] | | $2^\triangle$ |
| [=] | | $2^\triangle$ |
| [=] | | $2^\triangle$ |
| [=] | | $2^\triangle$ |
| [=] | | $2^\triangle$ |
| [=] | | $2^\triangle$ |

1. What is the calculator doing each time [=] is pressed? _____

_____

2. What number will be displayed next? _____

3. What pattern do you observe in the units digits of the powers of 2?

_____

4. What do you think the units digit of $2^{10}$ will be? _____

5. What is $2^{10}$? _____

6. What is the largest power of 2 the calculator is able to display? _____

7. Which is greater $2^{14}$ or $4^7$? _____ Why? _____

_____

8. What is the largest power of 4 the calculator is able to display? _____

## Thinking Cap

If you know the value of $2^{26}$ how can you find the value of $2^{25}$? $2^{27}$?

The Casio SL-450: A Tool for Teaching Mathematics

# TEACHER NOTES: *Power Up!*

**Objective:** To find powers of given numbers.

**Grade Level:** 6

**Topic:** *Exponents/Patterns*

## Using the Activity:

This activity uses the automatic constant for multiplication to find powers of given numbers. By having students enter 2 x x 1 to activate the automatic constant, a direct correlation is created between the number of times the equal key is pressed and the power to which the number is raised.

2 [x] [x] 1 [=]           gives 2 or $2^1$
2 [x] [x] 1 [=] [=]        gives 4 or $2^2$ equating to 2 x 2
2 [x] [x] 1 [=] [=] [=]     gives 8 or $2^3$ equating to 2 x 2 x 2

**Answers:**   2, $2^1$;   4, $2^2$;   8, $2^3$;   16, $2^4$;   32, $2^5$;   64, $2^6$;   128, $2^7$;   256, $2^8$

1. It is multiplying the number in the display by 2.
2. 512
3. The units digits repeat 2, 4, 8, 6, 2, 4, 8, 6, ...
4. answers will vary
5. 1024
6. $2^{26}$
7. They are equal. Each time you have two factors of 2, that is the same as having one factor of 4.
8. $4^{13}$

## Thinking Cap

This section has students focus on working from the known to determine the unknown. Students must understand that the previous power, $2^{25}$, is one-half the value of $2^{26}$, while the succeeding value, $2^{27}$, is double the value of $2^{26}$.

## Extension

Allow students to explore powers of other numbers to determine patterns and relationships. Repeat the activity with powers of 3 and 9. Offer reversal questions such as $\Delta^5 = 243$ or $\Delta^5 = $ **3125** or $\Delta^5 = $ **100,000**.

# Exit Sign

Name_____

Enter 9 into the calculator.

Press [+/-]. What is displayed?_____

Press [+/-]. What is displayed?_____

What does the [+/-] key on the calculator do?_____

Enter each exercise into the calculator. Record the sum displayed.

| Enter | Display |
|---|---|
| 1.  19 + 35 = | |
| 2. −19 + −35 = | |
| 3. −19 + 35 = | |
| 4.  19 + −35 = | |
| 5. −19 + 19 = | |

Study the displays.

How are the first two examples the same?_____

How did the calculator compute the sums?_____

How are the last three examples different from the first two examples?
_____

Were the answers computed in the same way?_____

How did the calculator compute the last sums?_____

According to your observations, what do you think the sums for the following exercises are? Verify your answers with your calculator.

| | Prediction | Calculator Answer |
|---|---|---|
| 6. −32 + −58 = | | |
| 7.  32 + −58 = | | |

## Thinking Cap

Use your calculator to determine the missing numbers in each display.

1. $27 + {-38} + \Delta = 8$, $\Delta = ?$   2. $-83 + \Delta + {-62} = 18$, $\Delta = ?$

© 1993 CASIO, INC.         **The Casio SL-450: A Tool for Teaching Mathematics**

# TEACHER NOTES: *Exit Sign*

**Objective:** To find the sums of integers.

**Grade Level:** 5-6

**Topic:** *Integers*

## Using the Activity:

The first part of the activity teaches students how to enter an integer into the calculator. To enter –9, press 9 [+/–]. The [+/–] key is the change of sign key and changes a positive number to a negative number and vice versa.

The second part of the activity is designed to have students discover the rule for adding integers. Students are expected to use number sense and reasoning to complete the activity.

Have students find the answer to exercises 1-5 on their calculator. Have them compare results and answer the questions.

*Answers:*

**1.** 54   **2.** –54   **3.** 16   **4.** –16   **5.** 0

How are the first two examples the same? *The numbers being added in each example have the same sign.*

How were the sums computed? *The numbers were added and sum has the same sign as the addends.*

How are the last three examples different from the first two examples? *The numbers being added have different signs.*

Were the sum computed in the same way? *No.*

How were the last sums computed? *The number parts (ignore the sign) were subtracted. The sum has the sign of the number further from zero. In the last example, since the numbers were equally far from zero, the answer is 0.*

**6.** –90   **7.** –26

You may wish to give students additional exercises to practice application of the rule.

## Thinking Cap

These exercises focus on number sense, along with addition of integers. Ask students to look at the sum given and the addends provided. Based on the sign of the answer, ask students to make predictions about the sign of the missing addend. For example, in the second problem, students should recognize the addend must be positive.

**1.** $\Delta = 19$   **2.** $\Delta = 163$

# Follow the Leader    Name_____

For each problem figure out by what number the calculator is counting on or counting back. Give the next three numbers in the pattern.

1. 1, 3, 5, 7, ___, ___, ___, . . .

2. 3, 9, 15, 21, ___, ___, ___, . . .

3. 8, 18, 28, 38, ___, ___, ___, . . .

4. 2, 13, 24, 35, ___, ___, ___, . . .

5. 4, 8, 12, 16, ___, ___, ___, . . .

6. 50, 48, 46, 44, ___, ___, ___, . . .

7. 34, 29, 24, 19, ___, ___, ___, . . .

8. 91, 82, 73, 64, ___, ___, ___, . . .

9. $.99, $1.05, $1.11, $1.17, ___, ___, ___, . . .

10. $1.28, $1.23, $1.18, $1.13, ___, ___, ___, . . .

Start with the given number. Make up a problem that involves counting on or counting back. You give 4 terms and have a friend give 3 more terms.

11. 38, ___, ___, ___, ___,

12. 19, ___, ___, ___, ___,

## Thinking Cap

Find the missing numbers in each sequence. Use your calculator to help determine the number used to count on or count back.

2, ___, ___, 17, ___, . . .

25, ___, ___, 13, ___, . . .

# TEACHER NOTES: *Follow the Leader*

**Objective:** To recognize and extend patterns.

**Grade Level:** 1-3

**Topic:** *Patterns*

**Using the Activity:**

Students are expected to determine the next three terms in a given sequence by discovering the relationship that exists between the terms in the sequence. Students must understand that whatever rule they determine for the sequence, must work for the entire sequence. They cannot change the rule in the middle of the sequence. Use of the calculator frees the student from the computation and allows the student to focus on finding the pattern.

Each sequence can be generated by use of the automatic constant. Students must first determine the relationship of the terms before they can key in the constant. For example, in problem 1, you are adding 2 to each term beginning with 1, so enter 2 [+] [+] 1 [=] [=] [=] [=] . . . In problem 4, you are adding 11 to each term beginning with 2, so you enter 11 [+] [+] 2 [=] [=] [=] . . . Sequences 1-5 involve counting on, while sequences 6 - 8 involve counting back. Sequences 9 and 10 involve money notation in a counting on and counting back situation respectively. Be sure to have students verbalize their thought process as they try to determine the rule.

*Answers*

**1.** 9, 11, 13; +2   **2.** 27, 33, 39; +6   **3.** 48, 58, 68; +10   **4.** 46, 57, 68; +11
**5.** 20, 24, 28; +4   **6.** 42, 40, 38; –2   **7.** 14, 9 , 4; –5   **8.** 55, 46, 37; –9
**9.** $1.23, $1.29, $1.35; +$.06   **10.** $1.08, $1.03, $.98; –$.05

Problems 11 and 12 are open-ended questions. Students should generate any sequence they wish beginning with the given number. The student should provide 4 terms of the sequence and then ask a partner to extend the sequence for 3 more terms. Compare sequences.

**Thinking Cap**

This section presents the student with a reversal of the first part of the activity. The student is given the first and fourth terms in each given sequence. Students must determine the value of the second, third, and fifth terms in each sequence. Students can use the guess and check strategy to determine the constant added in the first sequence and the constant subtracted in the second sequence.

Answers:  2, 7, 12, 17, 22 (add 5);   25, 21, 17, 13, 9 (subtract 4)

# Next In Line

Name_____

The sequence 3, 4.5, 6, 7.5, 9, ... was formed by adding 1.5 to each term to get the next term. Any two consecutive terms in this sequence differ by 1.5.
(4.5 − 3 = 1.5, 6 − 4.5 = 1.5).
In a sequence, if consecutive terms differ by the same amount, the sequence is called an **arithmetic sequence**. This sequence could be generated on the calculator by entering:

1.5 [+] [+] 3 [=] [=] [=] ...

The sequence 4, 8, 16, 32, 64, ... was formed by multiplying each term by 2 to get the next term. Any two consecutive terms in this sequence have the same ratio of 2
(8/4 = 2, 16/8 = 2, 32/16 = 2).
In a sequence, if consecutive terms have a common ratio, the sequence is called a **geometric sequence**. This sequence could be generated on the calculator by entering:

2 [x] [x] 4 [=] [=] [=] ...

**Use your calculator to determine if the following sequences are arithmetic or geometric or neither. Give the next 3 terms for each.**

1. 1, 2.5, 4, 5.5, _____, _____, _____, ...
2. 0.55, 0.8, 1.05, 1.3, _____, _____, _____, ...
3. 15, 14.25, 13.5, 12.75, _____, _____, _____, ...
4. 256, 128, 64, 32, _____, _____, _____, ...
5. 3, 7, 15, 31, _____, _____, _____, ...
6. 3, 9, 27, 81, _____, _____, _____, ...
7. 3, 64, 6, 32, 9, 16, 12, _____, _____, _____, ...
8. Use your calculator to generate an arithmetic sequence that starts with 1.5. List at least 4 terms in the sequence.
   1.5, _____, _____, _____, _____, ...
9. Use your calculator to generate a geometric sequence that starts with 4. List at least 4 terms in the sequence.
   4, _____, _____, _____, _____, ...

## Thinking Cap

36 is the fifth term of an arithmetic sequence and 56 is the ninth term of the same sequence. What is the first term of the sequence?

# TEACHER NOTES: Next In Line

**Objective:** To classify sequences and generate additional terms.

**Grade Level:** 4-6

**Topic:** *Patterns/Problem Solving*

## Using the Activity:

In this activity, students will use the automatic constant for +, −, and × to generate additional terms in a sequence and will classify the sequence as arithmetic, geometric, or neither. *Arithmetic sequences* have a common difference between consecutive terms. In generating the sequence on the calculator using the automatic constant for addition you would enter: **the value of the common difference** [+] [+] **first term in the sequence** [=] [=] [=]. Note, if the sequence is descending, going from higher values to lower values, but still arithmetic you enter: **the value of the common difference** [−] [−] **first term in the sequence** [=] [=] [=]. *Geometric sequences* have a common ratio (or quotient) between consecutive terms. In generating the sequence on the calculator using the automatic constant for multiplication you would enter: **the value of the common ratio** [x] [x] **first term in the sequence** [=] [=] [=]. The value of the common ratio could be a whole number or a decimal number. If in comparing consecutive terms in a sequence, no common difference or ratio can be found, then the sequence is neither arithmetic nor geometric. Students must determine the pattern by analyzing the relationship of consecutive and non-consecutive terms in the sequence. For example in problem 1, each succeeding term is 1.5 greater than the prior term, therefore the sequence is arithmetic generated by 1.5 [+] [+] 1 [=] [=] [=]. In problem 4, each succeeding term is 1/2 of the previous term, hence the sequence is geometric generated by .5 [x] [x] 256 [=] [=] [=]. In problem 5, there is no common difference or ratio. The difference between the terms changes from 4 to 8 to 16, suggesting the next terms is 63, a difference of 32.

*Answers:*    1. arithmetic, 7, 8.5, 10    2. arithmetic, 1.55, 1.8, 2.05
3. arithmetic, 12, 11.25, 10.5    4. geometric, 16, 8, 4    5. neither, 63, 191, 447
6. geometric, 243, 729, 2187    7. neither, 8, 15, 4    8. and 9. answers will vary

## Thinking Cap

This section explores relationships in an arithmetic sequence. To get from the 5th term to the 9th term the same number would have been added 4 times. The difference between the 2 terms is 20. Dividing 20 by 4 gives the constant difference of 5. Starting by subtracting 5 from 36 and working backwards, gives a first term of 16.

# Short Cut

Name_____

Use your calculator to find each product.

1. 15 x 15 _____
2. 25 x 25 _____
3. 35 x 35 _____
4. 45 x 45 _____
5. 55 x 55 _____
6. 65 x 65 _____

7. What similarities do you observe in each multiplication problem?
   _____

8. Look at the products and the factors you multiplied. Can you see a pattern? _____

9. Predict what 75 x 75 will be based on the pattern. _____ Verify your answer on your calculator.

10. Describe how the relationship between the factors and products above can be used to allow you to find the products mentally.
    _____

**Use your rule to find the following products. Verify your answer with your calculator.**

11. 105 x 105 _____
12. 125 x 125 _____

## Thinking Cap

Use your calculator to find the following products: a) 83 x 87  b) 94 x 96  c) 72 x 78  d) 46 x 44. Analyze the factors used and the resulting product. Explain how you can mentally find the product of two two-digit numbers that have the same tens digit and whose units digits sum to 10.

# TEACHER NOTES: *Short Cut*

**Objective:** To use patterns to determine a strategy to compute given products mentally.

**Grade Level:** 5-6

**Topic:** *Patterns/Problem Solving*

## Using the Activity:

The focus of this activity is to use the calculator to determine certain products that students will then analyze to uncover relationships that can lead to the development of a mental math-strategy for finding the products. The activity encourages the investigation and exploration of mathematics.

When multiplying a number that ends in 5 by itself, the product will always end in 25. Furthermore, if you take the digit that precedes the 5 in the factor and multiply that number by the next consecutive counting number, you will form a product that is the first part of the product sought. For example, in 25 x 25, the 2 precedes the 5. Multiply 2 by the next consecutive counting number, 3, to get 6. The final product can be thought of as having two parts; the first part of 6 and the second part of 25. Hence 25 x 25 = 625. For 65 x 65, think 6 x 7 = 42 for the first part and 25 for the second part, to get 65 x 65 = 4225.

*Answers:*  **1.** *225*  **2.** *625*  **3.** *1225*  **4.** *2025*  **5.** *3025*  **6.** *4225*
**7.** *In each problem, both factors are the same. All the factors end in 5.*
**8.** *The products end in 25. The first part of each product is the product of the number that precedes the 5 in the factor times the next consecutive counting number.*
**9.** *7 x 8 = 56 so 5625*   **10.** *The product of a number that ends in 5 times itself is formed by multiplying the number that precedes the 5 in the factor times the next consecutive counting number. That product is the first part of the final product and the second part of the product is 25.*
**11.** *11025*   **12.** *15625*

## Thinking Cap

The numbers explored in this section have the relationship of the factors being two two-digit numbers with the same tens digit and whose units digits sum to 10. The resulting product is found by multiplying the digit in the tens place times the next higher counting number and placing after it the product of the units digits of the given factors.   Answers: 7221, 9024, 5616, and 2024.

# Family Affair

Name_____

The calculator can be used to find percents of a number.
To find **50% of 60 ENTER 60 [x] 50 [%]**
The calculator displays 30.

**Use your calculator to find:**

1. 5% of 240 _____
2. 10% of 120 _____
3. 20% of 60 _____
4. 40% of 30 _____
5. What do you notice about the answers? _____
6. Explain why this relationship exists. _____

If you want to find different percents of the same number, you can use the constant function to speed up the the calculations.

**Enter the following to find different percents of 80.**

        80 [x] [x] 100 [%]
                    90 [%]
                    75 [%]
                    50 [%]
                    25 [%]
                    20 [%]
                    10 [%]
                     1 [%]
                   150 [%]
                   200 [%]

7. How is 20% of 80 related to 10% of 80? _____
8. How is 5% of 80 related to 10% of 80? _____
9. How could you find 75% of 80 if you knew 25% of 80? _____
10. How is 150% of 80 related to 100% of 80? _____
11. How could you find 200% of 80 if you knew 100% of 80? _____
12. How could you find 10% of 80 without a calculator? _____

## Thinking Cap

Use your calculator to find 95% of 60 and 60% of 95. What do you observe? Explain why.

# TEACHER NOTES: *Family Affair*

**Objective:** To find the percent of a number.

**Grade Level:** 5-6

**Topic:** *Percent*

## Using the Activity:

The focus of this activity is to instruct students on how to use the calculator to find the percent of a number. Typically, students change the percent to either a decimal or fraction, and then multiply that number times the given base to determine the percent. The calculator has the capacity to directly compute the percent of a number. To find **a% of b enter b [x] a [%]** and the calculator will automatically display the percentage.

In addition to learning how to use the calculator to find the percent of a number, the activity strives to develop in students a number sense about percent. Students are first asked to find percents of given numbers and then are asked to analyze these answers to determine relationships. Exercises 1-4 all have the same answer. Students should realize that this occurs since each time a percent was doubled, the base was cut in half. Anytime one factor in a multiplication is doubled and the other is halved, the product remains the same. Knowing this enables students to work from known facts to find unknown ones mentally. Questions 7-12 further develop percent relationships by analyzing the results of finding different percents of 80. The students use the automatic constant (80 [x] [x] ) to rapidly find the different percents. Use of the constant limits the number of keystrokes that must be entered for each problem. Students can discover many relationships such as 100% of 80 is 80; 50% of 80 is half of 80; 25% of 80 is one quarter of 80; and so forth. Understanding these relationships will help students to gain a stronger sense of percent, enabling them to make better estimates of answers to percent problems and do more problems mentally.

*Answers:* **1-4.** 12  **5.** all the same.  **6.** Each time the percent was doubled the base was cut in half resulting in the same product.
Percents of 80: 80, 72, 60, 40, 20, 8, 4, 0.8, 120, 160
**7.** double it  **8.** half of it  **9.** multiply by 3  **10.** one and half times 100%
**11.** multiply by 2  **12.** divide 80 by 10 by moving the decimal point one place left.

## Thinking Cap

This section teaches students an important property of percent: a% of b = b% of a. Students should see that (.95)(60) is the same as (.60)(95).

# Making the Grade  Name_____

Problem 1:

On a math test, Joe received 88 points out of a possible 100 points. What percent of the test did Joe get correct?_____

Would you need a calculator to figure out the percent?_____Why or why not?
_____

Mrs. Figures' math tests do not always total 100 points. However, she always reports her students' scores as percents. Mrs. Figures figures out the percents for the test using a calculator. If a test has 40 questions and a student gets 28 questions correct, Mrs. Figures presses:

$$28 \; \boxed{\div} \; 40 \; \boxed{\%}$$

The calculator displays the percent, 70.

**Use your calculator to figure out the percent scores on the following tests.**

| Name of Student | No. of questions on the test | No. of questions correct | Percent correct |
|---|---|---|---|
| 2. Spencer | 40 | 32 | |
| 3. Francisco | 40 | 24 | |
| 4. Jennifer | 40 | 16 | |
| 5. Samir | 40 | 35 | |
| 6. Elizabeth | 40 | 39 | |
| 7. Tasha | 20 | 16 | |
| 8. Hakeem | 52 | 39 | |
| 9. Michicko | 52 | 48 | |

## Thinking Cap

A school's scoring scale is:

90%-100% A;  80%-89% B;  70%-79% C;  60%-69% D;  below 60% F

If a test has 48 questions, how many questions must a student get correct to score at least a B?_____

# TEACHER NOTES: *Making the Grade*

**Objective:** To change a ratio to a percent.

**Grade Level:** 5-6

**Topic:** *Percent*

## Using the Activity:

The focus of this activity is to instruct students how to use the calculator to change a ratio or fraction to a percent. Typically, students are instructed to change the ratio to a decimal through division and then convert the decimal to a percent by multiplying by 100. Students could use the calculator to duplicate this process. However, the calculator has the capacity to convert a ratio directly to a percent. To convert **a/b to a percent, you enter a** $\div$ **b %**. This key sequence tells the calculator to find the quotient of a ÷ b and convert that quotient to a percent. In addition to learning how to use the calculator to convert a ratio to a percent, the activity provides a forum to help students learn about the relationship between ratio and percent. In the first problem, it is expected that students understand if you have a total of 100, by definition, the percent is obvious. In the situation where the total is other than 100, the process outlined in the Mrs. Figures scenario is followed. Have students analyze the numbers in the chart to recognize that 32 out of 40 is an equivalent ratio to 16 out of 20, and hence, the ratios have the same percent. Also since 16 out of 40 is one-half of 32 out of 40, the resulting percent is one-half of the other. Students should be made aware that if ratios are reduced to more commonly recognized ratios, the percent equivalent may be more obvious. For example, by reducing 39 out of 52 to 3 out of 4 the answer of 75% becomes obvious.

In converting ratios to percents on the calculator, often times the percent is given to multiple decimal places, i.e. 92.307692. Students should be instructed to round their answers.

Answers: 1. 88%, no, total is 100.  2. Spencer 80%  3. Francisco 60%
4. Jennifer 40%  5. Samir 87.5%  6. Elizabeth 97.5%  7. Tasha 80%
8. Hakeem 75%  9. Michicko 92.3%

## Thinking Cap

This section provides a reversal question regarding the relationship of a ratio and percent. Students must determine how many questions correct out of 48 will result in a percent equal to or greater than 80%. Give students hints such as "How do you know it has to be more than 24?" Students ought to know that 24 out of 48 is only 50%, hence it must be greater. Students can use guess and check to find the answer, 39.

# The Price is Right    Name_____

Many times at a store, the final selling price needs to be adjusted. Prices are often adjusted up, by the addition of sales tax, or adjusted down, by the subtraction of a discount. Calculators can be used to find the final price. There are two methods that can be used.

**SALES TAX:** A watch sells for $60. The store must charge 6% state sales tax. What is the final cost of the watch?

**Method One:** Think about adding 6% onto the original price.

**Enter: 60 [x] 6 [%] [+]**

The calculator displays 3.6 (the tax) when [%] is pressed and the final price of 63.6 when [+] is pressed. Since the answer is money, it would be recorded as $63.60.

**Method Two:** Think about paying 100% of the cost of the watch plus an additional 6% or 106% of $60.

**Enter: 60 [x] 106 [%]**

The calculator displays 63.6.

**DISCOUNT:** A dress sells for $45. The store has a 15% off sale. What is the final cost of the dress?

**Method One:** Think about subtracting 15% of the original price.

**Enter: 45 [x] 15 [%] [−]**

The calculator displays 6.75 (the discount) when [%] is pressed and 38.25, the cost, when [−] is pressed.

**Method Two:** Think about paying 100% of the cost less 15% of the cost or 85% of the cost.

**Enter: 45 [x] 85 [%]**

The calculator displays 38.25.

**Use your calculator to find the final cost of each item.**

1. A television sells for $280 plus 6% sales tax. Final cost_____
2. The bill for dinner at a restaurant is $48 plus a 15% tip. Final cost_____
3. A shirt sells for $24 before a 20% discount. Final cost _____
4. Which is a better buy: a $480 television offered at a 30% discount or a $420 television offered at 20% discount?

# Thinking Cap

Is a discount of 15% followed by a discount of 10% equal to a single discount of 25%?

# TEACHER NOTES: The Price is Right

**Objective:** To compute add-ons and discounts on prices.

**Grade Level:** 6

**Topic:** *Percent*

## Using the Activity:

The focus of this activity is to instruct students on how to use the calculator to find add-ons or mark-ups and discounts on prices. Typically, students would find the given percent of a price and in the case of an add-on, add it to the original price, or in the case of a discount, subtract it from the original price. The activity offers students two different ways to handle the add-on and discount situation on the calculator. The first method is:

price [×] rate [%] [+] for add-on
price [×] rate [%] [−] for discount

The sequence has the calculator compute the actual add-on or discount when the percent key is pressed and automatically add it or subtract it from the price when either the [+] or [−] is pressed.

The second method involves the reasoning that, if a price is marked up or added on to, you are paying the entire price (100%) plus an additional percent. The two percents are added together, and that percent of the price is computed. Conversely, for discount, you are paying less than the total cost (100%); therefore, the two percents are subtracted and that percent of the price is computed.

When dealing with finding percent of money, the calculator will often display answers to more than two decimal places. Students should represent answers to two decimal places.

*Answers:*   1. $296.80   2. $55.20   3. $19.20   4. *same price*

## Thinking Cap

In this section, students explore successive discount - discounts on items already discounted. A common mistake students make is to add the percents together and find that percent of the price for the answer. This gives a different result than first taking 15% of a price and then 10% of the resulting price. Suggest to students that they should actually select a price, such as $40, and discount first by 15% and then take 10% of that result. Next, they should take 25% of 40 and compare.

## Extension

Explore beginning with $25, marking the price up 20%, and then discounting the resulting price 20%. How does the result compare to the original price of $25?

# Square Deal

Name_____

The area of a square is found by multiplying the length of a side by itself.
Area of the given square = 14 units x 14 units = 196 square units

14

14

**Find the area of a square with given side measures.**

| side measure | area |
|---|---|
| 1. 28 units | |
| 2. 36 units | |
| 3. 44 units | |

If you know the area of the square and need to find the dimension of the side, the $\sqrt{\phantom{x}}$ on the calculator can be used. Square root ($\sqrt{\phantom{x}}$) tells you what number multiplied by itself gives the indicated value. Since the sides of a square have equal measure, the area is the product of a number times itself.

If the area of a square is 441 square units, what is the length of the side?

**Enter: 441 $\sqrt{\phantom{x}}$**    The calculator displays 21.

A square with side measure 21 units has an area of 441 square units.

**Use your calculator to find the length of a side of a square with given area.**

| Area | Measure of a side |
|---|---|
| 4. 361 square units | |
| 5. 1024 square units | |
| 6. 5329 square units | |
| 7. 2916 square units | |
| 8. 200 square units | |
| 9. 156.25 square units | |

# Thinking Cap

If the area of square A = 1 square unit and the area of square B = 4 square units, find the area of square C and the length of the side of square C.

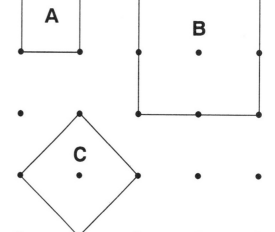

© 1993 CASIO, INC.

**The Casio SL-450: A Tool for Teaching Mathematics** 59

# TEACHER NOTES: *Square Deal*

**Objective:** To use the square root key to find the measure of the side of a square of given area.

**Grade Level:** 5-6

**Topic:** *Geometry and Measurement*

## Using the Activity:

The focus of this activity is to have students learn how to use the square root key on the calculator to find the length of a side of a square of given area. The activity begins with a review of how to find the area of a square. The formula typically used is $A = s^2$, where s is the measure of a side of the square. Since the SL-450 doesn't have a **$x^2$** key, the area is computed by multiplying the side measure times itself.

The second part of the activity reverses the process, investigating the use of the $\sqrt{}$ key. Finding the square root of a number is the inverse process to squaring a number. Hence, since the area represents the square of the side measure, to find the side measure you take the square root of the area. After finding the square root of the area, students should square the number to verify that the two processes are inverses of each other.

There are many numbers that students ought to be able to find the square root of without having to use the calculator. For example, the square root of 49 is 7, since 7 x 7 = 49. Ask students to tell you the square roots of 16, 25, 64, and 81 without using the calculator.

Answers:  **1.** *784 square units;*   **2.** *1296 square units;*   **3.** *1936 square units;*
**4.** *19 units;*   **5.** *32 units;*   **6.** *73 units;*   **7.** *54 units;*   **8.** *14.142135 units or 14.1 units*
**9.** *12.5 units*

## Thinking Cap

In this section, students should recognize that the area of C is greater than the area of A, but less than the area of B. Hence the area of C must be between 1 and 4. Visually, if the area of C is subdivided into 4 triangular regions by introducing the diagonals, the students should see that the 4 resulting regions can be combined to form 2 unit squares. Hence, the area is 2 square units. Using the $\sqrt{}$ on the calculator, it follows that the length of the side to the nearest hundredth is 1.41 units.

**60**   The Casio SL-450: A Tool for Teaching Mathematics